Dinner From Scratch

How to Raise Meat Chickens

BRIAN CUNNINGHAM

BRIAN CUNNINGHAM

Dinner From Scratch: How to Raise Meat Chickens by Brian Cunningham

Visit the authors' website at
https://ferrinbrookfarm.wordpress.com

Copyright © 2018 Brian Cunningham

All rights reserved. No portion of this book may be reproduced in any form without permission from the publisher, except as permitted by U.S. copyright law. For permissions, contact:

ferrinbrookfarm@gmail.com

ISBN:9781728701073

AUTHOR'S NOTE

Everything in this book comes from several years of experience working with hundreds upon hundreds of chickens first hand. Processes and plans are original and have been developed and refined as I learned from successes and failures. The concept of a moveable chicken pen or "chicken tractor" was not invented by me, and I thank the many clever farmers whose poultry housing I have observed and gleaned ideas from.

BRIAN CUNNINGHAM

TABLE OF CONTENTS

1 - WHY? 1

2 - COMMON MISCONCEPTIONS ABOUT RAISING CHICKENS FOR MEAT 2

3 - REFLECTION: THE FIRST KILL IS HARDEST ...5

4 - GENERAL CRITERIA TO RAISE MEAT CHICKENS...8

    IS IT LEGAL? 8
    LAND & SPACE REQUIREMENTS 9
    WASTE CONSIDERATIONS 10
    CONSIDERATION FOR NEIGHBORS 11
    CLIMATE & TIME OF YEAR 11

5 - SELECTING BREEDS 13

    STANDARD BROILERS (KNOWN AS CORNISH CROSS/ROCK)14
    RANGERS.. 15
    DUAL PURPOSE BREEDS.. 16

6 - HOW MANY CHICKS SHOULD YOU ORDER? ..19

    HOW MANY CHICKENS CAN YOU HANDLE RAISING? 19
    HOW MANY CHICKENS CAN YOU FIT IN YOUR FREEZER? .20
    HOW MANY CHICKENS DO YOU ACTUALLY NEED? 21
    DON'T ORDER YET—PLAN FOR DEATH 22
    ORDERING ONLINE VS. LOCAL STORE PICKUP..................... 22

7 - BUILDING 24

    PREPARING A BROODING SPACE 24
    General Brooder Supply Checklist: 24
    PLANS: ADAPTABLE PLYWOOD BROODER (FOR UP TO 12 BIRDS) 29
    BUILDING MOVEABLE CHICKEN PENS 33
    PLANS: MOVABLE CHICKEN PEN (FOR UP TO 12 BIRDS)...34
    HOW TO USE A MOVABLE CHICKEN PEN.................... 37

How Many Chickens Can You Fit In A Movable Pen? 37

8 - FEEDING: HOW OFTEN & HOW MUCH 39

Full Time Access To Food 40
12 Hours On, 12 Hours Off 40
Two Restricted 30 Minute Feeding Windows 41
How Much Feed Will You Need? 44

9 - RAISING THE BIRDS 45

Day 1 46
Brooding supply checklist: 46
Weeks 1 - 2: In The Brooder 47
Daily Checklist for Weeks 1-2: 47
Weeks 3 - 4: In The Brooder 48
Daily Checklist for Weeks 3-4: 49
Day 29: Introducing Chickens To Pasture 49
Pasturing supply checklist 49
Catching loose chickens! 51
Weeks 5 And On: In The Moveable Pens 52
Daily Checklist for Weeks 5 and on: 53
The Day Before Slaughter 54
When To Cull 54
Bird is not moving very much and won't stand to eat or drink 54
Bird is limping or has crooked toes 55
Purple comb 55
Gasping for air 55
Deformities 55
Deciding to eat a cull 56
Unexpected Mortalities 56
Disposing Of Birds You Will Not Eat 57

10 - PROCESSING (SLAUGHTER) 58

Processing Day Supply List: 59
Setting Up For Processing 60
The Kill 63

Killing with a restraint cone 63
Killing with an axe and old feed bag 64
SCALDING 65
PLUCKING 66
EVISCERATION 66
PACKAGING & FREEZING 69
CALCULATING COST PER POUND 70

11 -   MASTER SUPPLY LIST 72

1 - WHY?

Because you know you will make the right choices. The chicken you eat will have been raised the way you want it to be raised. Exactly. You choose its feed, you choose its home, you choose how it is treated. You keep them healthy and content.

Because you love food. The texture will be exquisite. The taste will be unique to you and your land.

Because you'll also develop a deeper connection with your humanity as a meat-eater. You will directly connect with your meals. You will earn your food. You will be a real provider.

Because you will truly make your dinner from scratch.

The real question you need to ask if you have the land and resources is: why not?

2 - COMMON MISCONCEPTIONS ABOUT RAISING CHICKENS FOR MEAT

If you've lived your life on chicken from the grocery store, I have some bad news: you have never tasted a true chicken. The depth of flavor achieved by a local, organically raised bird will astound you, as will the tender meat that is so much harder to overcook than a factory sealed bird. The smell of a true chicken slowly roasting in the oven will drive you mad with delight and anticipation.

There are a lot of theories and ideas on how to raise chickens out there, and a lot of what you'll find is based on fear and misconceptions. Before you get too freaked out, here are some of the major myths floating around out there that you should take with a grain of salt.

1. Meat chickens are especially stinky.
WRONG. Believe it or not, all animals have to poop. Rarely does it smell good! The intense stink associated with chickens, especially those raised in larger quantities, is always a result of poor management. If you move your chickens regularly (as with movable pens) or follow a

routine to clean their living space, the smell will never become an issue.

2. Raising chickens for meat is not cost effective.
WRONG. This is only true if you are interested in buying the cheapest chicken available in the grocery store. But then again, are those even really chickens? You're not interested in raising a dry, flavorless bird, anyway, you want to eat true chicken. Even though feed costs have increased recently, the net cost is still a bargain considering the end results. Without doing too much math, the cost per pound of organic meat can be as low as half the cost at a grocery store. You'll get to have full control over humane treatment and get to say thank you.

3. Meat chickens inevitably have health issues.
WRONG. Virtually all meat and produce we consume has been bred in ways to emphasize traits that are favorable, and there is no denying that poultry breeders have selected birds for their ability to efficiently convert feed to meat. Cornish cross breeds are especially good at packing on meat and as a result they really go to town on their food and can develop some strange issues. But from our research and personal experience, even the crazy Cornish Cross can thrive if given the opportunity. While it's true this breed can be more prone to some issues, it is almost always going to be the result of something that can be improved by altering management of the flock. Overfeeding is the chief concern, and is easily avoided with some attention. We have raised hundreds of broilers and dozens of hens—all in all, broilers aren't much different from our lively hens when treated right. Treat them like happy birds and they'll follow suit!

4. Meat Chickens require too much work.
WRONG. We like to use moveable pens, so there is virtually no cleaning work required. We just let them

fertilize the land beneath their movable pens for a day, then move the pen to a fresh spot. It worked wonders on our lawn. Other than the daily moving of the pens, the only other daily work is feeding (two controlled feedings a day) and making sure they always have water. This works out to less than 10 minutes twice a day. If you can't spend 20 minutes on some of the best meat around, you will just have to miss out. As for the final processing, you will find that if you plan it out conservatively, have some helpers, and don't attempt to process too many at once, it will be smooth sailing. We'll talk more about processing in a later post.

5. Meat Chickens are too violent.

WRONG. We have observed some aggression with our egg laying hens, mostly in the form of normal pecking to establish dominance. But we find that the meat birds seem to get along great, more apt to spend time cuddling in the shade than to hurt each other. Since most meat breeds reach market weight between 6 and 14 weeks and we keep them separate from our older birds, they never get a chance to learn aggressive habits.

The most common thing I hear about these locally raised chickens, far more common than any of the above, is how darn good they taste. Or I have I mentioned that already? You can totally do this!

3 - REFLECTION: THE FIRST KILL IS HARDEST

Trigger warning: I don't want to come off as a psychopath or anything in this book. But there is no way around it—if you want to raise chickens for meat, you (or someone you hire) will need to kill the animal you raised. Killing anything living is very difficult, and may not be for everyone. If you don't think you can handle it, or don't think you can find someone to do it for you, please reconsider this endeavor, and maybe don't even read this.

Still here? Thanks. It's hard but worth it. I'm writing this as an outlet for reflection about a life-changing experience. I'm sharing it with the intention of giving full disclosure in the context of this book, but really I want to share so we can all have some support and give and receive empathy for something very intense and difficult.

Here's how it went for me.

When you are first pursuing a project or hobby, there are two ways to go about it. One is to figure it out as you go along, and the other is to do as much research as possible, stock up on all the necessary gear, and feel like you left no stone unturned. Maybe we're more into the research and prep and gear these days, but for our first flock of meat chickens, we were largely figuring things out as we went along. Good for learning and finding a system that works for us as individuals, bad for knowing what you're getting into.

Instead of investing in restraint cones to position the bird perfectly to start the process of slaughter, I opted to cut the corner out of a feed bag large enough for a chicken's head and neck to poke through. I read about this in a forum.

Instead of a sharp, precise knife to make a clean cut and let the bird bleed out in a controlled manner, I used an axe. Like a goddamned Thanksgiving cartoon.

I caught the chicken and positioned it in the bag on a log. There was surprisingly little struggle, for whatever reason. Maybe my slow, nervous movements translated to something peaceful in the bird.

I picked up the axe, holding the bird in the bag in place with my foot. I wondered how high to raise the axe for the best aim and impact. I raised my arms and a flash of nervous energy filled every inch of my body.

I stepped back. My heart was pounding, and I felt dizzy. I paced in a circle and the bird stood up, bag and all. I could see only its head, flicking side to side as it tried to figure out what was going on.

This really was like a cartoon.

I took a deep breath. We couldn't keep these chickens as

pets. We weren't going to spend money to call in a professional. There was no turning back. This had to be done. And it had to be done right then.

I took another breath and got the bird and myself back into position. I apologized to the bird.

There was one last moment when the chicken was alive and calm on that stump. Then, like jumping into cold water, I slammed the axe down as hard as I could. This was followed by an explosion of movement within that bag. Feathers and blood sputtered. On the ground, the chicken bounced out of the bag and flopped down the hill a few feet. After an unknown number of seconds or minutes, all was still.

I had made a clean cut right where I needed to in the middle of the neck. The bird probably didn't suffer. I hope not. It's hard to ever really feel sure about that.

After it stopped moving, I could still feel every blood cell in my veins. The worst was over, but in that moment I felt the worst of all as it all sunk in completely.

After a short break, I got through four more birds that day. I wouldn't say it got easier, but the intensity of that first kill was never matched.

It is hard to participate in the food chain sometimes, but I do feel more connected to what it means to be a human and to earn one's food, having been through this experience. For putting myself through it.

We cooked that first chicken in tomato sauce served with pasta for some friends a few weeks later. The taste was utterly profound.

4 - GENERAL CRITERIA TO RAISE MEAT CHICKENS

If you aren't sure whether you can or should raise chickens, here are some factors you will need to answer before diving in:

Is It Legal?

Sometimes even when you own the land, there are covenants or laws that will prevent you from raising poultry (and/or other animals). As a first stop, our suggestion would be to reach out to your local cooperative extension. They will either know the answers right away or may be able to point you in the right direction. If not, check in with your town/city officials. You don't want to make a big investment in a project where animal lives are involved and find out you can't legally follow through, so don't skip this!

If raising chickens is legal where you live, but you can't slaughter and process them, you may still be able to raise them and have someone else process them. Check with local butchers, or maybe you can find a service that makes "house calls" and can process legally on-site with mobile equipment.

Land & Space Requirements

At a bare minimum, you will need 2-3 square feet of living space per chicken. You will want to take odor and health safety into consideration here. The location of the living space should not be near water sources (check local laws and restrictions to see how close livestock/poultry can be to a well, for example). You also want to keep yourself and your family clean, so don't locate the chickens in a spot that will be regularly used for other purposes, like a patch of lawn near a swing set.

If you are using moveable pens, you will need to do some math to determine how much space you'll need. For example, let's consider a 5'x6' pen. In order to move the pens every day for 3 weeks, you will need to have 21 5'x6' patches, or 630 square feet. This is slightly more than a 25' x 25' patch of land. You will also need to consider the fact that all of this land will be getting trod and pooped upon by curious chickens that will dig and peck around. You have to be willing to let this land get a little roughed up for the duration, and then some. We've found that lawns take about a month to get back to normal—well not normal, since the grass is definitely greener after a flock of chickens has passed over it! There are occasionally some bare spots leftover, but new seed will perk right up into fresh grass thanks to all the fertilizer.

Waste Considerations

There are two forms of waste you will encounter when raising chickens for meat. The poop (let's say… manure) and the waste of processing (feathers, entrails, etc.).

Manure can be managed either by using movable pens and letting nature do its thing, or you can collect and compost. As you may suspect, chicken manure is excellent compost and shouldn't be thrown away!

As for the waste from processing, you'll have to be thoughtful. There are at least the following options:

1. Garbage - make sure it's legal to dispose of stuff like animal remains (heads, feathers, blood) before considering this. And even if it is legal, keep in mind that you will need to take extra steps to bag things up and prevent leaks and contamination. We don't recommend throwing processing waste away.
2. Composting - you can compost much of the waste, from feathers to blood and everything in between. However, this time of compost fodder will almost definitely attract local predators, or even neighborhood pets. From mice to coyotes, you will have visitors, and they may not be so neat and tidy as they browse. You can cover the fresh material with grass, pine shavings, household compostable waste, etc. to reduce attracting wildlife. But don't expect to keep everything away. And be sure to follow good composting practices and regularly mix the pile and ensure everything has thoroughly composted before using. Composting is also somewhat problematic and should be done with great care.
3. Burying - again, if you are sure it is legal to do so

on your property, you could dig a deep hole and bury the waste. We suggest digging deep enough to have 6-12 inches of earth covering the waste. This will help keep predators and local pets away.

Consideration For Neighbors

We think it is very important to be aware of your impact not just on the earth and on local wildlife, but also on your neighbors. Chickens can get noisy as they get close to market weights, with roosters even beginning to crow as early as the 7th week sometimes. And while some practices like using movable pens can greatly diminish odors, there can still be some stink when raising chickens for meat. If you have close neighbors, take this into consideration. You may even want to be respectful and let them know what you're up to so they understand what's going on and what to expect.

Climate & Time of Year

Just like pretty much any animal, chickens don't like it too hot or too cold. When it is too hot, chickens are actually at risk of death due to dehydration or simply overheating. Avoiding high heat—or managing it, is important with larger meat breeds, since they can be more susceptible to health issues even in good temperatures. Hotter chickens may also eat less, which means lower meat yields. Hey, you want to farm, you need to think about yields!

When it is too cold, we've noticed chickens tend to end up more fatty. And standard broilers don't put on as thick a coat of feathers as other breeds, which can leave them more exposed to the cold. Don't let temperatures get too low with this breed, even if they are mature.

The most important climate factor comes into play during brooding, the first 2-4 weeks. You will start the baby chicks at a consistent 95 degrees Fahrenheit—if you can't maintain that, you may need to look into raising the birds at a different time of year when you can.

Another factor to consider is processing day. You will likely be doing this work outdoors, or at least partially outdoors. Too hot or too cold can make this unpleasant.

As a general rule, we recommend trying to raise your birds between spring and mid fall. This can be suitable for most regions in the US. In New England, we've started birds as early as May and finished as late as November.

5 - SELECTING BREEDS

To be very general, there are essentially three different types of chicken that folks commonly raise these days: Egg layers, meat chickens, and hobby chickens. This book will solely focus on meat chickens—we're trying to make our dinner, after all! But it's important to know that the image of a chicken you have in your head could be very, very different from the chicken that you will raise for meat. You might think all chickens are generally the same, but brace yourself. The fastest growing, most efficient meat chickens grow fewer feathers and can get somewhat ugly as they transition from cute yellow chick to a fully feathered meat beast!

Of the various breeds you can raise for meat, there are a few subcategories to pick from:
- **Standard broiler**, also known as Cornish cross/Cornish rock
- **Rangers**
- **Dual Purpose** breeds (dual = egg layers and decent meat production)

Let's cover the different stats and factors of each of these to help you determine which breed is best for you.

Standard Broilers (Known as Cornish Cross/Rock)

This is the type of chicken you'll find in most grocery stores. When you get chicken for under a dollar a pound, it's a safe bet this is the type of chicken.

Pros:
The reason for their popularity is a combination of cost, efficiency, and easy logistics. They grow their meat quickly, reaching a desirable market weight in 6-8 weeks, typically. That's right, two months or less from chick to dinner-ready! And the amount of food they need to consume in order to reach market weight is much more efficient than any other breed, so they put on more muscle faster without the farmer having to buy as much feed as with other types of chicken.

They don't need much space, either. See a comparison of space requirements and other factors at the end of this section.

Broilers are also bred to have white feathers that pluck easily—this means the end result will look cleaner and more appealing. Even if some small feathers or stems of feathers are left behind on the bird, the white color keeps things looking good.

Cons:
Broilers are more prone to having health issues. Sometimes they grow too quickly for their own good, and can't support their own weight without injury. Or they can

simply eat too much and get sick.

Another factor that we notice is that when broilers are raised for maximum yield and profit, it can be unpleasant. The birds are not very birdlike when they are constantly stuffing themselves and don't live in a nice setting. They will lie around most of the day and only show interest in food. This can be overcome with some practices we cover in this book, but even still, don't go into this expecting broilers to be as fun to watch as a nice laying flock.

Rangers

There are a few options available to the backyard chicken farmer that don't grow quite as quickly as the standard broiler. These go by some different names, but generally are referred to as "Rangers" of some kind, since their breeding has given them more interest in foraging, or "ranging" around for food.

Pros:
While they aren't as fast growing as a common broiler, they still grow quickly and put on a lot of meat. The general window for these birds to reach a market weight is 9 to 11 weeks.

The slower growth can encourage healthier fats to develop on the bird, and if you free range or pasture them, the nutritional potential may also increase. And with more time growing, the flavor of these birds also increases—many believe that Rangers have superior meat to common broilers.

Another point worth making is that a flock of rangers is much more enjoyable to be around. They are more like hens in that way—they don't lie around all day, they cluck, they explore. Watching chickens is my favorite part of raising them. It's relaxing and helps clear my head.

Cons:
The slower growth of Rangers means you have to invest more time and energy into raising these chickens. You'll also spend more money on feed, as you'll have additional weeks to feed them and they don't convert that feed to meat quite as efficiently as common broilers.

Rangers usually have darker feathers (commonly red, like a Rhode Island Red's feathers), which looks great while they are out on pasture, but could potentially become problematic if plucking doesn't go smoothly. Any feathers or pin feathers left behind will be very obvious, so you'll need to watch out for that.

Rangers will also need some more space than a common broiler to satisfy their needs for foraging and generally higher levels of activity. See table at the end of this section for details.

Dual Purpose Breeds

Chickens that are dual purpose would include many of the common egg-laying breeds, such as Buff Orpington, Barred Rock, and Jersey Giants. Heritage breeds that are dual purpose would be the ultimate way to get as close as possible to what chicken used to be like, before breeding techniques got intense. There is usually information right on the order forms/websites that show the differences between the breeds, so I won't bother getting into the different breeds and options out there in this book. The differences are not going to be that profound with these breeds, for the most part.

Pros:

The main reason to raise a dual purpose breed is that it can get you can experience something that is kind of rare these days. You may find that you truly enjoy it, as we do. The meat requires different cooking technique and will likely lead you to cook some dishes you never considered making before. Slow cooked recipes are ideal. The flavor and nutritional value can also be much deeper with these birds, as they have had more time to mature.

You can also order "straight runs" of chicks, and get a mix of hens and roosters to raise. This would be cheaper per chick, and will get you some hens to keep for eggs.

Cons:

Dual purpose breeds take much longer to raise, at least twice as long as a common broiler, clocking in at 16 weeks at least before they reach any sort of marketable weight. This means more time, effort, and expense. They will also need more space as they grow (see chart at the end of this section). It is also worth noting that, by sixteen weeks, you'll probably have roosters crowing at 5am (and 6am, and 7am, and 2pm, and 6pm!), which can get on one's nerves easily.

Plucking these breeds can also be potentially problematic, as they will have fully developed plumage by processing time, which will be sturdier and likely darker colored, so you may be in for some extra plucking effort to get to clean end results.

Quick Comparison of Meat Chicken Breeds:

	Time to Market Weight	Feed Conversion Performance	Can you keep females as layers?	Additional Notes
Standard Broiler (Cornish Cross, Cornish Rock, Cornish X)	7-9 weeks	Excellent	No	Can be processed at 3-4 weeks for Cornish game hens
Rangers	9-12 weeks	Good	Yes	Excellent foragers, nice looking when feathered
Dual Purpose & Heritage Breeds	16-20+ weeks	Bad	Yes	Example breeds: Buff Orpington, NH Red, Barred Rock

Note that, for the purposes of simplicity and wider appeal, the information in this book largely applies to raising standard broilers (also known as Cornish cross, Cornish rock, Cornish x). Other breeds will have similar requirements/information, but note that there could be differences that are not called out.

6 - HOW MANY CHICKS SHOULD YOU ORDER?

Here are some obvious statements that are so obvious, maybe you'll be like we were when we first started and not bother to even consider them:

- How many chickens can we handle raising?
- How many chickens do we actually need?
- How many chickens can we fit in the freezer?

Important things to think about! Let's take a quick moment to walk through these.

How Many Chickens Can You Handle Raising?

Know your limitations. First, how much space do you have for chickens? Each standard broiler needs a minimum of 2 square feet in their living space (we aim for at least 3). If you are using movable pens, do you have enough land to move the birds on a daily basis? If you need help understanding this, see the Land & Space Requirements section earlier for more info.

The other limitations that you can't forget about are time and effort. When things are going smoothly, you probably only need 15-30 minutes of effort a day to move pens and maintain food and water. This may seem minimal, but don't forget—it's every day for 7+ weeks straight, no skipping a day! And then you need to think about processing time and effort (if you are processing the birds yourself). How many days or hours are you capable of giving up for processing chickens? Depending on a variety of factors, it can take between 5 and 30 minutes to process a single bird. You don't need to process all of the birds in one day, but make sure you do the math here and make sure you don't bite off more than you can chew. We tend to aim for processing between 10 and 25 birds on a single day, and don't make any other plans that day just in case.

How Many Chickens Can You Fit In Your Freezer?

We recommend you get a chest freezer if you don't have one already. These days they are pretty efficient and don't cost a whole heck of a lot. And you can increase your capacity for a year's supply of meat, potentially!

It's obvious—you need to factor space into how many chickens you raise. But how do you figure it out? How many cubic feet does a chicken take up? As a general rule, we recommend planning on the following:

- Calculate a total by assuming 6 birds per cubic foot of freezer space
- Subtract 10% from total number of birds to play it safe

With the above you will probably have leftover space, but if you ask us, it's better to have the extra space rather than not knowing what to do with the last few birds! And don't forget that these calculations assume an empty freezer to begin with—if you are storing other food you will need to take that lost space into account.

How Many Chickens Do You Actually Need?

Got a handle on the two factors above? Before you place an order, take just a few more moments and consider how much meat you actually need. You don't want to fill up your freezer only to realize you have a 3 year supply. Here are a few factors to think about that should help you get a sense of how many birds you'll actually need:

- Think in terms of calendar. How many chickens would you eat per month (or week)?
- Will you sell any chickens? If so, we recommend getting orders secured before you order chicks (this means you need to get orders about 2-3 months before the meat is actually ready).
- Will you give any away? Family and friends like chicken.
- In addition to the above, will you want any birds for special occasions? Thanksgiving chicken ain't that bad!

Don't Order Yet—Plan For Death

OK, by now I think you should probably have a number in mind. But a final point to consider: chickens sometimes die before you can process them. Sickness, predators, escape—you name it. Even with excellent farming practices, chickens can be susceptible to these means of early departure, and it's important to factor it in.

Here's our rule: Plan for 20% to not make it, but be able to support it if they all survive.

So when you order, if you have the resources and can make the time and effort for all of them, we recommend increasing your total by up to 25%.

Ordering Online Vs. Local Store Pickup

You can order chicks in a few ways, but the main options available to most folks (and the two ways we've done it) are ordering from a local farm supply store and ordering online.

Farm stores will either have chicks in stock that you can pick from, usually in the spring, or you can fill out an order form for specific ordering windows (check with the store for details). When the chicks are ready, you can pick them up in the store and bring them right home. Getting chicks from a store will most likely be cheaper because you don't have to pay for your individual shipping costs.

Ordering online offers a lot more flexibility and selection. Online retailers will have calendars for each breed so you can determine exactly what you want and when you get it, based upon availability. These retailers are able to offer many more breeds than farm stores, typically. And if you dig around, you can find smaller scale hatcheries or even individual farmers offering chicks or eggs and you can get to know how they raise their chickens. But you will have to pay shipping costs, which may not be too cheap, so factor that in. In our experience, once the chicks arrive at the post office, you'll get a call to come pick them up. It's a fun trip!

7 - BUILDING

In order to keep your chickens safe and healthy, they're going to need a place to live. There are two types of housing you'll need to prepare if you follow our practices:
- Brooder for chicks
- Moveable pens

This chapter will walk you through the requirements of these two types of housing so everything can go as smoothly as possible. And of course, there are detailed plans for building your very own!

Preparing A Brooding Space

General Brooder Supply Checklist:
- ☐ Space for brooding that meets these requirements (like a shed or garage):

- Secure from predators and pests
  - No wind
  - Ventilation / windows
  - Easy access for care and cleaning
  - Can hang heat lamps up to 4 feet high
  - Adaptable from small to large space for heat efficiency is a plus
  - 1 square foot per 2 chicks, plus space for feeders and waterers
- Heat lamps
- Chain to hang heat lamps
- S hooks for heat lamps
- Newspapers
- Pine shavings
- Bricks to raise feeder/waterer over time, or more chain and s hooks
- A good supply of broiler starter feed

Baby chicks are sensitive, almost helpless creatures. But their list of needs is relatively short:
- Warm, consistent temperature
- Adequate space
- Shelter that is secure from predators and prevents their accidental escape, and is easy for you to clean

Raising them in these conditions is called brooding. A mother hen could brood her chicks and fulfill the above needs, but I doubt you have a willing hen to cover a sudden flock of chicks.

Temperature
The common way to maintain temperature for brooding chicks is to use a heat lamp. You can get a lamp specially designed for the purpose of brooding at most farm supply stores or online. The lamp is really just two pieces—the

fixture itself and a special heat bulb. The fixture will have a clamp to help with mounting, but otherwise it's just a light socket with a bowl over it to direct the heat.

We always hang our lamps with a chain and S hooks. Hanging makes it so it's unlikely the clamps will fail and the lamp goes crashing to the ground. And chain with S hooks makes it easy to adjust the height of the lamp—something you'll need to do to hit just the right temperature in the brooding space.

Here is a quick guide to the temperatures you need to maintain, week by week:

- Week 1 (0-7 days): 95 degrees F
- Week 2 (8-14 days): 90 degrees F
- Week 3 (15-21 days): 85 degrees F
- Week 4 (22-28 days)—if birds still aren't fully feathered: 80 degrees F
- Week 5, or whenever the birds are fully feathered—no heating necessary

To control the temperature, make sure you have a brooder with walls to hold the heat. If it is especially cold, you may need to position some blankets or other barriers over the top to help hold the heat in.

For taking readings, we recommend using an instant read infrared thermometer to take temperature readings in a fast and reliable way. This kind of thermometer is available at home improvement stores, usually by the tools.

To set up the lamp, start by hanging it so it is about 18 inches from the floor of your brooder. You can't get much closer than this without risking overheating the chicks. Take temperature readings in the hottest part of the beam of light on the floor. If it's too warm, raise the lamp by a as needed until you hit the target temperature.

Note that as the seasons advance and temperatures rise or fall, you may need to adjust the lamp position to maintain the right temperature. Take readings multiple times a day during that first week to make sure you stay close to the right temperature.

Also note that, at night, temperatures will drop. Check on the temperature before you go to bed, and check again when you wake up, adjusting as needed.

TIP: Always have a backup heat lamp bulb ready to go in case the one in use burns out so you can keep your chicks warm!

Setting up or building a brooder
In the simplest terms, a brooder is just four walls with a lamp hanging over it. The key is to ensure it is secure and accessible.

Unless you want to build a new free-standing structure for brooding, we recommend building a brooder inside of an existing structure, like a garage, shed, or abandoned chicken coop. You could also brood in your house, but beware that it will be stinky and maintenance will be a little messy.

You will need to plan to have a brooder that is large enough to give each chick **1 square foot of space**. When they are smaller you could temporarily have less space, but they WILL grow. You will also need a place for waterers (sometimes called "drinkers") and feeders. The product details on the waterers and feeders will specify how many birds they can support.

Tip on selecting waterers: We prefer the cheaper plastic 1 gallon waterers because we've found that they are about as durable as any other waterers and are easy to multiply for larger flocks. In our experience, metal waterers tend to get rusty and/or fail after about 2 years, which is frustrating considering their higher cost.

Do some math and figure out how many square feet you need, and then you can map out your rectangle. Or if you have a shed or chicken pen that approximately matches the dimensions you need, feel free to use those, as long as you can access for cleaning and care.

For the first week, the chicks will need to be on newspaper, rather than pine shavings. This is because curious chicks might eat the pine shavings and end up dying because they can't properly eat food with a piece of shaving stuck. The paper will be messy, and you'll probably have to change it 2 times a day to keep up with the mess. This might sound rough, but remember that it's just for a week—pine shavings are much easier to manage for the rest of the brooding, and then the birds will be out on pasture fertilizing your land, no need to clean!

Then you can upgrade to pine shavings. Spread enough for the birds to comfortably walk around, maybe a 1-2 inch depth at first, and you can increase as they get taller. You'll want to replace the pine shavings 1-2 times per week, depending on how many birds you have and how much space you give them.

PLANS: Adaptable Plywood Brooder (For Up To 12 Birds)

Feel free to get creative with how you want to build a brooder, but if you want some specific direction, this is a smaller version of what we do. We build this in a detached garage on a plywood floor, so if you are using these plans you will need to start with a structure that is safe from predators and weather, with a flat floor. We leave the top of the brooder open for easy access, but do carefully cover some of the open top with an old blanket if it gets especially cold, keeping the blanket from touching the lamp.

When you are finished brooding, take everything apart and stack neatly. It's almost like there were never any chickens!

Note on flocks larger than 12 birds: these plans can easily be modified to accommodate more birds by adjusting the size of the plywood walls to suit your space requirements. Note that for larger spaces, additional materials may be required—more feeders, waterers, and heat lamps.

Tools
- ☐ Saw to cut plywood
- ☐ Hammer

Materials
- ☐ Secure space for brooding, like a freestanding shed or empty chicken coop
- ☐ 1x 4'x8' sheet of plywood for every ~14 birds
- ☐ 12 concrete blocks (standard size, 16" x 8" x 8", hole pattern up to you)
- ☐ 6 standard bricks or 3 1'x1' pavers per waterer or feeder

- ☐ Heat lamps
- ☐ Chain to hang heat lamps
    - Length depends on height of space you are using
    - Most types of chain will be sufficient for this purpose, as long as you can use S hooks with it
- ☐ S hooks that will fit the chain you use
- ☐ 3" common nail
- ☐ Plenty of newspaper
- ☐ Pine shavings (will be used after first 2 weeks)

Process

1. Sweep the space (shed, garage, etc.) you will be using and double check to ensure it is secure from predators and will stay dry.

2. Lay a double layer of newspaper to cover a 4'x4' area.

3. Cut the plywood: You will want to cut the board into 4 equal pieces that are 4' long and 2' high, so one rip cut down the middle, and then cut the two halves into two:

DINNER FROM SCRATCH: HOW TO RAISE MEAT CHICKENS

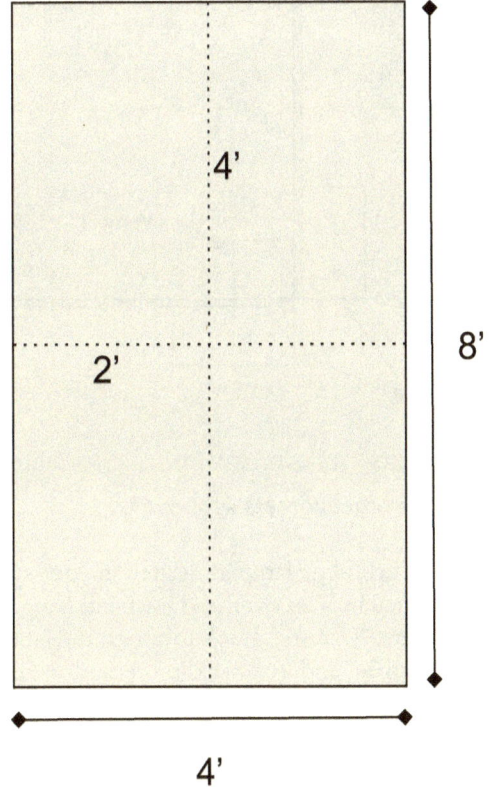

4. To set up the plywood walls, start with two pieces to make a corner—make sure the boards are positioned so they are 2 feet high and the walls are 4 feet long, and they meet at a 90 degree angle. On the inside corner, set up a concrete block vertically to sit right in the corner. Place two more concrete blocks on the outside edges of the corner, flush with the edges of the plywood.

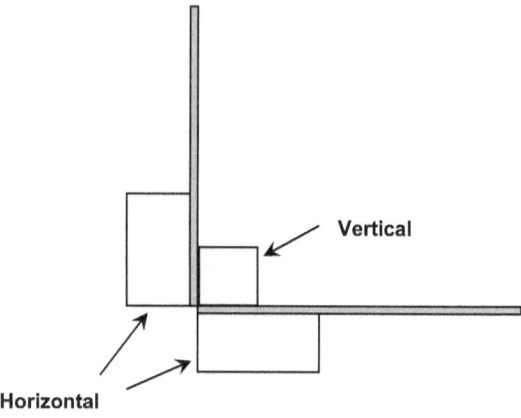

5. Set up the other corners as above.

6. Hang the brooding lamp as desired in your space—we hammer a nail into an overhead joist and hang the chain from there. Use an S hook to attach the lamp hanger to the chain, and as a backup precaution also attach to the chain directly using the lamp's clamp (if it has one). Adjust the height of the lamp as needed to reach your target temperature (95 degrees Fahrenheit for starting baby chicks). You can loop excess chain onto the nail to keep it neat—don't let it hang into the chicken coop as it could interfere with the lamp.

7. Add food, water, and chicks! As the birds get taller, you can raise the height of the feeder(s) and waterer(s) using bricks or pavers.

8. Replace newspaper 1-2 times a day for 2 weeks, then switch to pine shavings.

Building Moveable Chicken Pens

We recommend raising meat chickens using moveable pens because they are efficient, offer great security, and allow you to control where your chickens are making their beautiful messes.

Maybe you've heard of the "chicken tractor", but if you haven't, the concept is extremely simple: it is essentially a cage with an open bottom that you keep your chickens in so they can graze on the land and get fresh air and natural light. It enables a "free-range" kind of solution that ensures your chickens are completely protected from predators.

And to keep things clean and sanitary, instead of enduring the miserable job of cleaning out a stinky coop and run every day from a multitude of birds, you can just move the pen every day to a fresh patch of land. The chickens will be happy, and the stink will be minimal, since the droppings are spread out and not concentrated. And actually, you'll be adding great fertilizer to your land. We regularly run our pens over different sections of lawn and the grass always recovers nicely, with a lush green boost from the fertilizer.

PLANS: Movable Chicken Pen (For Up To 12 Birds)

Tools
- ☐ Aviation snips or similar to cut hardware cloth
- ☐ Work gloves
- ☐ Hacksaw or similar to cut PVC pipe

Materials
- ☐ 8x 10-foot lengths 1" PVC pipe
- ☐ 4x 1" PVC tee fittings
- ☐ 4x 1" PVC elbow fittings
- ☐ 8x 1" PVC side outlet elbow (corner) fittings
- ☐ 6' x 8' Tarp for roof
- ☐ 50 nylon cable ties, at least 5" long
- ☐ 8 blocks to pin down bottom flaps (could also stake these down or use logs or rocks)
- ☐ Around 20' of clothesline or similar rope
- ☐ At least 32' of 36" tall hardware cloth (½" holes are good)
    > *Note: Chicken wire is not recommended, since some predators are able to reach through the larger holes and grab chickens and kill them.*

Procedure
1. Carefully measure and cut the PVC pipes according to the below cut list using hacksaw. The 28" lengths can be cut from the scraps of the 6' cuts.
    a. 6x - 5' lengths (from 3 pipes)
    b. 6x - 3' lengths (from 2 pipes)
    c. 5x - 2' lengths (from 1 pipe)
    d. 2x - 6' lengths (from 2 pipes)
    e. 2x - 28" lengths (from remaining 4' scraps)

DINNER FROM SCRATCH: HOW TO RAISE MEAT CHICKENS

2. Using 2 each of the 5' and 6' lengths and 4 of the side outlet elbows/corners, assemble the bottom frame, which will be a 5'x6' rectangle. Make sure all side outlet elbows point up. Install 4 of the 2' lengths into the tops of each corner.

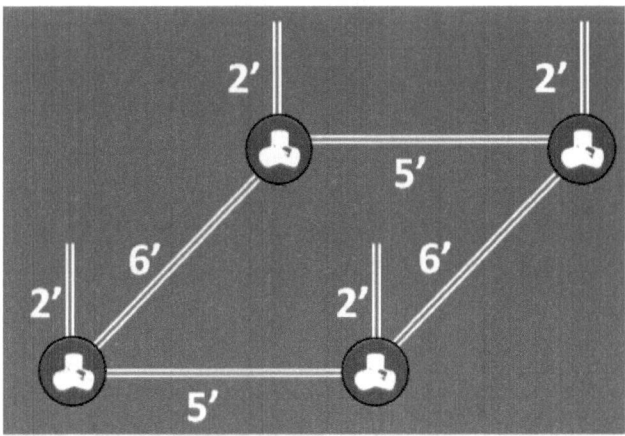

3. Assemble top frame as shown. A crossbar for hanging food & water is made with the 28" pieces supported by the 2' piece & tee foot. Attach top to bottom via 2' lengths.

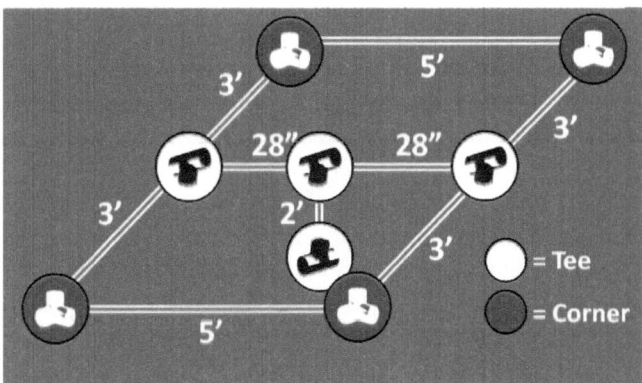

4. Assemble door frame as shown, using elbows, 3' lengths and remaining 5' lengths. Cover door with hardware cloth cut to size and attached with ties. Attach to crossbar of main structure using loosely tied rope to allow door to open and close easily.

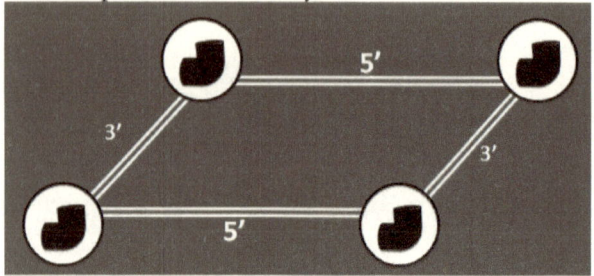

5. Flip structure and attach one end of hardware cloth to a corner with cable ties. Make sure the wire extends upward by 1 foot. Continue attaching by wrapping around the frame and secure with ties at the corners and along the frame. Cut hardware cloth at corners straight down to the frame to create flaps. Cut away excess.

6. Flip frame upright and fold out the flaps at the bottom. Cover remaining opening at top of frame with hardware cloth, securing with cable ties and overlapping at seam in the middle. Fold at overlap to eliminate gap. Cover with tarp for a roof, securing with rope where convenient and making sure to cover the food and some of the sides to reduce exposure.

7. Pin down the flaps using logs or blocks to prevent predator access from the bottom. Use rope or chain to hang food and water from center support. Tie loops of rope around frame to make pulling the pen easier. Bring on the chickens!

How To Use A Movable Chicken Pen

When your chicks have feathered, they are ready for going outside into the movable pen. Once they are in, the process is so simple, it only takes 2 steps:
1. Start the pen from one end or a corner of your open land.

2. Every day, move the pen far enough for it to entirely cover a new patch of land.
Of course, you'll also want to make sure to feed and water the chickens per your management plan.

How Many Chickens Can You Fit In A Movable Pen?

The guidelines for this vary, but we've seen a fairly consistent number of 2 square feet per meat chicken at minimum. We give ours plenty more space when we use movable pens, but from my observations of the chicken behavior and the frequent cuddling that goes on, this guideline is probably OK.

So, how many chickens you can fit depends on the size of your pen. You can either plan your pens to fit a certain number of chickens, or figure out your maximum based on the dimensions of your pen, like so:

1. Take the length, multiply by the width to get the total area of the pen

2. Divide by 2 to give each chicken 2 square feet

3. Subtract one chicken from the result, since the feeder and waterer will take up around 1 square foot each (adjust this as needed for larger feeders/waterers).

General Tips
- Make sure the bottom is flat on the ground. Any gaps could allow sneaky predators in.
- Make sure to have ~1 foot of hardware cloth extend outward from bottom of pen to prevent digging predators. Consider weighing down the flaps if they do not lie flat.
- When you move the pen, go slow and be careful. Watch your birds and don't injure them if they aren't paying attention!

8 - FEEDING: HOW OFTEN & HOW MUCH

One year we conducted an experiment and raised two very different flocks of chicken. Both were raised on pasture in our moveable chicken pens, both were given organic feed to supplement that lush pasture, and both were the same breed from the same hatchery, the famously fast-growing Jumbo Cornish rock breed that makes the news a lot these days. Everything we did was the same for each flock except for one thing: how much time they had access to food every day.

Before we get into the comparison of our two flocks from the experiment, let's go over the basics of feeding broilers. Generally speaking, there are three options.

Full Time Access To Food

First-time farmers might naively feed their broilers like this. It is how most people feed hens. But hens are often heritage (or close to heritage) breeds that are resilient and were not bred to be eating machines. Unlike a broiler, a hen will hold back, and many breeds actually prefer to forage.

But listen, you cannot go with instinct when it comes to these intense, high growth breeds.
They are not like hens, and will positively stuff themselves with food, and come back for more after a short break. This can lead to health issues and higher rates of mortality.

12 Hours On, 12 Hours Off

This easy alternative to 24 hours of access is probably the most popular feed solution folks use with these fast-growth breeds. And this is because, as we've seen in our experience, fewer birds have health issues when they have their feed taken away for those 12 hours. So you have less loss.

And on top of that, the yield of meat per bird remains close to the yield from birds with 24 hours of access.

It's surprising this works so well, since you'd think most of those 12 hours would be at night anyway, when the birds can't possibly be stuffing themselves, but there are a few hours as the sun rises and sets that they are left to forage and relax, which makes all the difference. This is the economical solution that business-minded farmers can exploit for the most profit while reducing the cost that comes with wasted feed and dead birds.

Two Restricted 30 Minute Feeding Windows

This feeding practice goes beyond the 12 hours method above with two goals in mind:
- Further reduce risk of health issues
- Promote foraging of plants and bugs for healthier, tastier meat

In the morning, you bring the food out, then 30 minutes later you take it back. The birds take a break from eating, and then get to work on the plants and bugs in the vicinity. We raise our birds on what is essentially an overgrown lawn, so there is a lot for them to browse. And in our experience, birds raised like this not only stay healthier and are generally more active and lively, they also taste much, much better. And according to some research, the meat (and fats) of a properly pastured chicken are actually better for you.

The Numbers: How do these broiler feeding methods compare in practice?

For our experiment we raised one flock on the 12 hours on, 12 hours off practice, and the other flock was given 30 minutes of access 2x a day. And we collected data!

And lucky for you, we've been collecting data for all of our flocks! Over the years, we've tried all three of the feeding practices and are quite intrigued by the results.

The key numbers we'll pay attention to are:
- Mortality rates
- Average processed weight (after slaughter, plucking, and evisceration)

- Feed conversion (lbs feed to produce 1 lb of processed weight)

Please note that these numbers are based on our experience and our farming practices, and are not necessarily the same as what you'd experience and certainly not the same as what commercial farms experience. Other factors are at play, including weather, predators, and level of experience. But we think the differences between the feeding methods are clear and would likely carry over to other farms.

So without much further ado, here are the averages across all of our broilers raised to date:

Effects of Different Broiler Feeding practices
(averaged across several flocks)

Feeding Practice	Mortality rate	Avg processed weight	Feed conversion
24 hours access	20%	6.21	4.3
12 hours on, 12 hours off	7%	6.08	2.5
2x 30 min feedings	0%	4.99	2.1

Hey look at that! We had no loss in our flocks with the two separate feeding windows. That feels very, very good!

Clearly, the mortality rate in our experience has dropped right along with a reduction in time allowed to access feed.

This is great. Lower mortality means healthier birds. We can tell you first hand that they were more active, loved to forage, and were generally more enjoyable to watch. And we think a happy farmer is a better farmer, and better farmers raise better food.

You'll also notice that the average processed weight has gone down too. In our experience, you lose about 1 lb of processed weight per bird when you feed for about 30 minutes twice a day a day vs. 12 or 24 hours of access every day. Also note that this reduced yield is slightly counteracted by better feed conversion ratios, so the effect on yields may not hurt as much when you consider feed costs as well.

So which do you choose? Profit? Or better tasting, humane, and healthy?

What it all comes to is this: Are you in it to make money or are you trying to improve the quality of the path that food takes to get to the plate?

If you are in it for profit, you should probably go for the 12 hours on, 12 hours off method of feeding your broilers. You may experience some loss to health issues, but it will easily be made up for by increased meat yields.

If you want to have healthier birds that we think taste much, much better, the 2 feeding windows are the way to go.

I can tell you now that we're never going back to 12 on, 12 off. It's just not worth it when animals' day to day lives are affected by difficulty walking, general inactivity, and having little interest in foraging—an instinct that can be completely erased by feeding practices!

How Much Feed Will You Need?

Generally speaking, you can plan on needing 10-15 lbs of feed per broiler if you are raising them to 7 weeks. About ⅓ of that feed should be starter feed, and the rest can be grower feed. Other breeds will require less food to get to 7 weeks, but you will be raising them for much longer to get to a market weight, so your mileage may vary.

Unless you have a large operation or are planning multiple flocks through the year, we don't recommend buying all the feed you calculate you'll need at once. This is because you never know when disaster can strike. A tree could fall on your brooder, a fox could find a way in, or your birds could get sick out of nowhere. We tend to buy feed in 50 lb bags, about 4-8 bags at a time.

9 - RAISING THE BIRDS

In the simplest way of looking at it, raising chickens for meat is really just a matter of regularly checking to make sure the following needs are consistently met:
- ☐ Maintain correct temperature (only applies for first 3-4 weeks)
- ☐ Food supplied per feeding practices (this will change after first 2 weeks)
- ☐ Birds have enough space (this will change as they grow)
- ☐ Ensure safety from predators
- ☐ Protect against the elements
- ☐ Clean water is always available

If you can check off all of the above every time you check on your birds, you're good. But some things do need to change as the birds grow up, so let's break things down a bit.

Before you dive into the below, review the earlier sections of this book, which contain information on the preparation and planning you will need to do in advance of getting your chicks.

Day 1

Brooding supply checklist:
- ☐ Space for brooding
    - Secure from predators and pests
    - No wind
    - Ventilation / windows
    - Easy access for care and cleaning
    - Can hang heat lamps up to 4 feet high
    - Adaptable from small to large space for heat efficiency is a plus
    - 1 square foot per 2 chicks, plus space for feeders and waterers
- ☐ Heat lamps
- ☐ Chain to hang heat lamps
- ☐ S hooks for heat lamps
- ☐ Newspapers
- ☐ Pine shavings
- ☐ Bricks to raise feeder/waterer over time, or more chain and s hooks
- ☐ A good supply of broiler starter feed

After you pick up your chicks, you'll need to introduce them to the brooder. If you haven't set up your brooder yet, see the earlier section in this book, which includes plans for a basic brooder setup. And before transferring the chicks to their new home, take a few moments to make sure you have everything set up—you have enough waterers filled to support the volume of chicks you have, feeders are ready, and the temperature is staying around 95 degrees F. And I hope you didn't forget the newspaper!

Now, one by one, gently pick up the chicks and dip their beaks into some of the water. Try to make sure they get a gulp down. This will teach them where their water source is.

Once they are all in, make sure you close any doors or lids you have for your setup, and walk away. Let them calm down and get familiar with their new lives. They'll find the food on their own and will start chowing down in no time.

On the first day, check on them every few hours to make sure their food and water levels are doing OK and there haven't been any major spills.

Weeks 1 - 2: In The Brooder

For the first 2 weeks, you will need to check your birds 3-4 times a day. When you check them, simply run down this checklist and maintain as needed:

Daily Checklist for Weeks 1-2:
- ☐ Temperature is good (95 degrees F for the first week, 90 degrees for the second week).
- ☐ Broiler starter feed is always available
- ☐ Clean water is always available
- ☐ Replace soiled newspaper
- ☐ Observe the birds' health

A note on observing health: Identifying health issues is important, because it could be an indicator of a problem with your setup or care routine. If you find an unhealthy bird, review your brooder set up and if you need to make any changes, make them as soon as you can. A healthy chick will be moving around quickly and eat and drink without any problems. Note that they will kind of look like they're dead or dying while sleeping—if you are concerned you can gently wake them to observe. Any that seem to be overly tired could be sick and could be on their way to dying. All you can really do to help ailing birds is make sure they are getting water and ensure the temperature is still good. If any do end up dying—don't feel too bad. This is an unfortunate part of raising chickens and happens to everyone.

Weeks 3 - 4: In The Brooder

By the start of week 3, your birds are probably looking pretty weird as their feathers come in. They're probably also getting big. Here are some changes to make:
- If necessary, increase the size of your brooding space so they have 1 square foot per every bird.
- Switch from newspaper to pine shavings.
- Make sure you still have enough waterers and feeders to accommodate your flock. You may need to add more as they start crowding around the waterers/feeders.
- Stop leaving feed in the brooder all day. This will help them gain weight more slowly and safely. Our recommendation is to transition to 12 hours on, 12 hours off with feed. We put the food out first thing in the morning and take it away around dinnertime.

Once you have made the above changes, you can probably plan to check on the birds 3 times a day—once in the morning when you bring out the food, once midday, and once when you take their food away at night.

Daily Checklist for Weeks 3-4:
- ☐ Temperature is good (85 degrees F for the third week, 80 degrees F for the fourth week).
- ☐ Broiler starter feed is available during the day (taken away at night)
- ☐ Clean water is always available
- ☐ Check pine shavings for strong odor or visible soiling—replace twice a week
- ☐ Observe the birds' health

Note: Some flocks will feather faster than others. If your birds are fully feathered and the daytime temperatures are in the 80s, you may be able to bring your birds out to pasture after 3 weeks. Keep a close eye on the weather forecast if you want to do this, though—rainy weather or cold spells can be dangerous for young birds.

Note on Cornish Game Hens: At around 3 weeks, the birds will be the right size to be Cornish game hens, which would be between 1-2 lbs for a processed bird. We sometimes take the opportunity to cull any birds with questionable health at this point.

Day 29: Introducing Chickens To Pasture

Pasturing supply checklist
- ☐ Movable pen(s) sized to support your flock comfortably
- ☐ Tarp to cover around half of movable pen completely (all of back side, ⅔ of top, ⅔ of two sides)
- ☐ Rope or zip ties to secure tarp
- ☐ Extra logs or blocks to weigh down the movable pen's bottom flaps

- [ ] Feeder(s)
- [ ] Waterer(s)
- [ ] Hose that can reach pens
- [ ] Extra chain and S hooks
- [ ] A good supply of broiler grower feed

After 4 weeks your broilers should be fully feathered and ready for the world. Before you move them from the brooder, make sure to get your moveable pens set up completely. See earlier section on movable pens for more information and some plans. Roofing/tarps should be in place, water and feed should be full and hanging at the right heights. Getting everything ready will help the birds have a less stressful transition, since you can place them in their new home and then walk away without further interacting. Some of them will certainly be a little freaked out, and bustling around to set up a tarp or replace a feeder will keep them in a stressful state for longer.

And before you add birds, make sure your pen is in a good starting point. The obvious location would be a corner of the area you will be using as pasture, since this allows you to slowly and evenly work your way up and down the land in neat and tidy rows without having to backtrack to an open spot after you run the birds over half the land. And don't forget that you will eventually need to process these birds—if you can plan to have the pens end up close to the processing area after all of the moves, that will be nice and efficient and save you some extra effort on processing day.

See the diagram here for a general idea of what the path of the pen should look like.

DINNER FROM SCRATCH: HOW TO RAISE MEAT CHICKENS

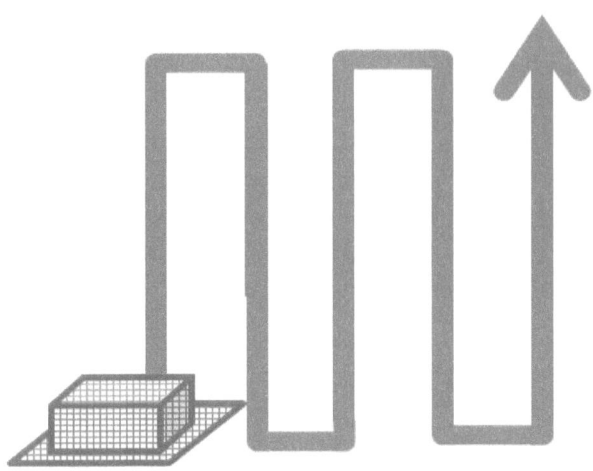

To move the birds, we recommend a large plastic tub with opaque sides that you can pick up on your own (or with a helper for larger tubs). Don't forget to account for the weight of the birds when you consider how many you can fit—they aren't light and fluffy little chicks anymore! A lid can also be helpful to keep the birds calm and prevent escape. Don't overcrowd them in the container—they may move around and you don't want them scratching each other up because there isn't enough space. Now it's just a matter of making as many trips from the brooder to the pen as necessary. Try to do all of this as gently as you can to avoid injury and unnecessary stress.

Catching loose chickens!
If you find yourself having trouble catching a bird, here are some tips:
- Try to get the bird into a corner or other location where they don't have as many escape options.
- Use your arms or even yard tools (like a leaf rake) to fan out your approach. You can guide a chicken in the direction you want by using your arms from a distance—for example, if you want

the bird to go to the right, stick your left arm out and aim it to the left of the bird, and the chicken will instinctively move away from your arm and go where you want it.
- When the time comes to grab the bird, be quick and assertive. Hesitation will result in failure. The easiest way to grab a bird is to go for their legs. These are easier to get a hold of and, as long as you aren't too frantic, it won't hurt the bird.
- Once you have a hold of the bird, pick them up by holding your hands over their wings. This will help calm them, and prevents them from flapping their wings around which could startle you into dropping them, or could injure their wings.

Weeks 5 And On: In The Moveable Pens

Once the chickens are in their pens, you are kind of on easy street. No more worrying about cleanup up newspaper or pine shavings.

Some changes to note in week 5:
- You will want to switch to broiler grower feed from now until the end. We use organic broiler crumble.
- You need to decide which feeding practice you will use. After raising hundreds of birds and trying various methods, we have decided that our birds are better off with two thirty minute feedings a day, once in the morning and once in the late afternoon. This not only keeps any health concerns with overeating at bay, it also encourages foraging. You can read more about this in the "Comparison of the 3 Common Broiler Feeding Practices" chapter of this book.

Daily Checklist for Weeks 5 and on:
- ☐ Move the pens once a day
- ☐ Make sure the covering is in place and has not been disturbed by wind or predators—birds need overhead protection from sun and rain at all times
- ☐ On very hot days, make sure sides of the pen are open to increase airflow
- ☐ On rainy days, make sure at least 2 sides of the pen are covered to provide shelter from wind and splattering water
- ☐ Broiler grower feed is available per your chosen feeding practice
- ☐ Clean water is always available and plentiful, especially on hot days
- ☐ Observe the birds' health

Notes on moving the pens
- You should try to move the pens exactly as far as you need to get the entire pen onto fresh pasture, and no further. This will make your use of the land more efficient. And try to use straight lines!
- You can move the pens at any time of day, but we prefer mornings because this gives them cleaner pasture to forage in for their waking hours.
- It is very important that you can see the birds as you move their pen. Remove the covering if you need to. Move the pens slowly, and watch to make sure you don't run over any legs—if you pull over a birds leg and don't notice, you could break their bones which can limit their ability to move around and eat and drink, not to mention cause them suffering.
- After the pen is in its new spot, make sure you replenish water and secure the pen in place with logs or rocks or stakes, whatever you've decided to use.

The Day Before Slaughter

You don't need to change anything for the birds before slaughtering EXCEPT for feeding. To make processing cleaner and easier, it is recommended that you make sure the birds have not eaten any feed in the 12 hours leading up to slaughter. You can just take away their food the night before. But make sure they have plenty of water!

When To Cull

As you raise chickens, you may run into a situation that requires intervention and killing a chicken. This is a hard truth. Culling is important for reducing suffering of a sick or injured bird, and it can save you a lot of stress and, to be frank, money spent on wasted feed. Here are scenarios we look out for when deciding whether to cull a bird or not.

As a general rule, if a bird is not interested in eating or drinking or is seriously injured, it is best to cull them in most situations. They will not likely get better and we consider it our duty to reduce suffering in animals when we raise them. Below are some specific scenarios to look out for.

Bird is not moving very much and won't stand to eat or drink
If a bird is just sitting down all day and you don't see it running towards the food when you bring it, this is a problem. They are most likely suffering from a leg problem and cannot put weight on their legs. It can also make the daily moves of the pen difficult, since you may not be able to get the bird to walk along with rest of the birds and the pen. A bird in this state will have difficulty surviving and putting on meat for processing day, and it is

best to cull them while they are still relatively OK so you can still use the meat.

Bird is limping or has crooked toes
Sometimes a bird with a limp will recover and continue to put on meat. If you see a bird limping, don't immediately jump to culling. Give it a few days and watch to see if it starts moving less and less. See above.

Purple comb
If a broiler has a purple comb, they could be sick. Watch them for other symptoms, like lack of movement, gasping, or a very swollen abdomen. If they have a purple comb and aren't moving very much, it is best to cull to save them from getting any worse. These birds can usually still be eaten, but we aren't veterinarians or USDA inspectors so you will have to research and decide on whether you want to eat your culls.

Gasping for air
Most chickens will pant or even gasp for air on very hot days. It can look scary, but usually when the night cools them off they are fine. If a bird is gasping for air and the struggle does not stop when things are cool, you could have a sick bird. Look for other symptoms like a purple comb or swollen abdomen. In this situation, you can usually tell that a bird needs to be culled. Trust your instinct. And a quick note: on hot days, make sure to uncover the sides of your pens to allow for better airflow.

Deformities
Sometimes you get a chick that has a deformity. We once had a chicken with a beak that didn't really come together correctly and was missing an eye. When you get a chick with a weird deformity like this, it is possible that they will end up dying due to the struggles with eating and drinking, but maybe not. We recommend observing any birds like

this closely—sometimes they will be OK and can make it to processing day along with the rest of the flock with minimal issues. If they are having trouble, you may need to cull.

Deciding to eat a cull
Generally speaking, if the meat and innards look OK, and the bird died at our hands, we will thoroughly cook it and eat it. If we find the bird dead or things look off with the meat or innards, we don't eat it.

But note that we are not veterinarians or USDA inspectors, so you will have to do your own research and decide on your own whether you want to eat a cull.

Unexpected Mortalities

We never eat a bird that died unexpectedly. This allows us to rule out any concerns with the meat sitting out for an unknown amount of time, and prevents us from consuming any unknown diseases.

If a predator killed a bird, you could potentially eat it if you recover the bird soon after its demise. But we have not done this, perhaps out of some kind of respect for the bird who went through a rough end. Plus we don't know much about where that predator has been and what germs it could be carrying.

Disposing Of Birds You Will Not Eat

If you have a dead animal, you need to check with what is legal on your land in your town/city. You might be able to bury them or you could throw them in the garbage. These kinds of laws are not universal so check with your local authorities.

Another interesting option is to offer the bird to the forest, which can sound tempting. But note that if you do this, you will attract predators to the area, and this might clue them into your other birds and maybe even your pets. It is pretty much never advisable to feed wild animals. And this may be illegal where you live. Check with local authorities.

10 - PROCESSING (SLAUGHTER)

Slaughtering birds will be a big deal. You are taking a life and you need to honor that. You must do everything you can to keep the bird calm and comfortable. You must reduce suffering. We will provide a framework for how we approach processing, but a warning: things will be difficult at first, at least emotionally.

But don't lose sight of what you have accomplished here: you are providing yourself and/or your family with meat that was treated the way you want it to be treated. You are a provider. You are participating in a part of humanity that has been lost in many ways with modern technology and corporate efficiencies. You should be proud. If you want to eat meat as part of your diet, this is an honest, direct, and engaging way to do it. And you are taking some business away from bad companies.

Thanks for coming on this journey with us.

Before you process, you have to collect supplies and set everything up.

Processing Day Supply List:
- [ ] Plucker (optional)—you can pluck by hand if you don't have a plucker, it just takes more time and effort
  - If using a plucker, make sure you have any additional supplies for the plucker setup. Here's what we use with our barrel style plucker:
    - [ ] Y splitter for garden hoses if you only have one spigot
    - [ ] An extra hose to connect to the plucker
    - [ ] Extension cord for plucker
- [ ] Garden hose that can reach your processing area
- [ ] Spray nozzle for garden hose
- [ ] Enough coolers to hold 10 birds
- [ ] Work table that can easily be washed and disinfected, like a foldable portable plastic table
- [ ] Trash barrel/bucket with trash bag
- [ ] Restraint cone (optional) mounted on something sturdy (a tree, saw horse, etc.)
  - If not using restraint you will need: chicken feed bag with corner cut out, an axe, and a large stump or wide log that you don't mind getting dirty or hitting with an axe.
- [ ] Bucket for draining chickens
- [ ] Rope for draining birds
- [ ] Large pot for scalding birds for plucking and shrinking bags
- [ ] Dish soap
- [ ] Candy thermometer
- [ ] Spray bleach solution
- [ ] Nitrile gloves
- [ ] Sharp paring knife or similar

- ☐ Plastic cutting board
- ☐ bowl
- ☐ Paper towels
- ☐ Work clothes and shoes that you aren't afraid to throw away if necessary
- ☐ Safety glasses
- ☐ Quart jars (optional, helpful for draining)
- ☐ Old towel or cloth you aren't afraid to throw away
- ☐ Bags to store chickens and any parts you are saving
- ☐ Source of entertainment (optional): we like to listen to podcasts while processing. In our opinion, listening to music is a bad idea—you might end up forever associating the songs you listen to with the experience of processing, and the next time you hear the song you might flash right back to the work you will be doing with your chickens.

Setting Up For Processing

When arranging your processing area, think of everything in phases. We arrange everything in a sort of line, starting from one end with live birds and ending up with cleaned and fully processed birds at the other end.

Also be conscious of where you set things up. You'll need to be able to reach the area with hoses and maybe extension cords, depending on your equipment.

And you will be cleaning up the area thoroughly, but the blood and feathers won't be completely eliminated and will likely attract animals, from small mice up to large predators, depending on where you are. You wouldn't want those animals right up against your house. You'll also want it to be a space that can be avoided for a while afterwards for sanitary purposes—you don't want to step in chicken feathers and then track them into your house, for example.

Take into consideration who might be able to see what you're doing. Neighbors could be upset to see processing, as would your kids. If you have kids, we recommend setting up in a place that they can easily avoid when they play.

Also select an area that is easy to clean when you are finished. And frankly, it should be a spot where a few feathers being left behind isn't a big problem. We do our work by the edge of the woods, away from our lawn and play areas.

Batch sizes and prepping for scalding

Also as part of your setup, you'll want to plan out how many birds you can kill, scald, and pluck in batches. You are mainly limited by the scalding water, which can be difficult to keep at the right temperature if you are heating it up on a stove like we do.

If you have a dedicated scalder, you will probably be able to move more quickly and might not even have to think in terms of batches. Follow the directions for your scalder to determine what process flow works for you.

You'll want to make sure to get the scalding water heating up on the stove before you kill any birds. If you have a standalone scalder, follow the directions that came with it.

Target 155-160 degrees Fahrenheit for scalding water.

For the way we work, we find that a 3 gallon pot of water heats up to about the right temperature if we set it over medium heat when we first start processing. By the time we get through 4 birds, the water is hot enough.

The number of birds you can scald at once will depend on your scalding vessel. We use a canning pot, which holds about 3 gallons of water. In this pot we can fit 2-3 broilers, depending on how big they are. To make the work move quicker, we use 2 pots like this at once and plan to scald 4 birds at a time. Those 4 birds fit comfortably in our plucker. And as mentioned above, we can get through 4 birds in the amount of time it takes for our scalding water to heat up. So our batch size is 4.

If you don't have a plucker, you may want to scald fewer birds at a time, since the work of removing feathers is time consuming.

Another factor in batch size is how much space you have in coolers for chilling birds after they are eviscerated. We have somewhat standard sized coolers that can accommodate 4-6 fully processed broilers comfortably.

To identify the ideal batch size for your setup and your desired style of work, start with 2 at a time and see if you can get up to 4, or maybe even 8.

Planning the day
Make sure you don't feed the birds in the morning of processing day—this can result in a mess when you are cleaning them up.

And don't bite off more than you can chew, especially if this is your first time processing. We recommend planning on processing 5 or 10 birds on your first day so you can identify any concerns with your supplies or setup. This will also help you determine how many birds you are comfortable working through in a given amount of time.

As a point of reference, we like to aim for processing 20 birds in a single day, which takes somewhere around 4-6 hours, including all set up and cleanup. This is with two adults working, usually in shifts, since we are also hanging out with our kids. Two adults could easily cut this time in half if they didn't take parenting breaks!

Something else to consider is help. If you get help, the work will go faster. You can offer payment in the form of meat!

OK, let's move on to the actual process.

The Kill

We plan out the movement of our pens so they end up close to our processing area, so we can just pick out chickens directly from the pen as we process. If you can't do this, we recommend using a large plastic tub and bringing a couple chickens over at a time for processing. Keep the work out of their sight, though. This is out of respect and will help them stay calm.

Killing with a restraint cone
1. Put on your gloves and safety glasses.
2. Check to ensure cone is securely mounted.
3. Gently guide the bird into it head first.
4. If necessary, gently tug the bird's head down through the bottom opening by holding their comb or gently holding their head with your index finger and thumb. They will usually be very calm at this point.

5. With the thumb and index finger of your subdominant hand, get a firm but not tight hold of their jaw and back of their head.
6. With your dominant hand, position the blade against the spot just under their jaw/beak, where it is usually very soft and some of their red waddle is. You may need to make sure there aren't any feathers in the way of the blade—these can interfere with the cut and end up only scratching the bird, stressing them and yourself out big time and requiring a second pass.
7. With a quick, assertive motion, pull the blade across this soft area and allow the blood to drain. If you did it right, the blood will be flowing out in a steady stream. If it is just pulsing out or dripping, you will unfortunately need to try again.
8. Allow the bird to drain until it is still. Once it is no longer moving, you can remove it from the cone and hang it by its feet using rope attached to a sawhorse, tree, etc. This will allow for further dripping and keep the bird clean before moving on to the next step.

Killing with an axe and old feed bag
1. Make sure your axe is ready and sharp and you have a stump or other firm surface that can handle a thwack from an axe, and some mess.
2. Prepare the feed bag by shaking out any crumbs and cutting a 3 inch opening at one of the bottom corners.
3. Guide the bird into the bag so its head and neck stick out at the end.
4. Position the bird onto the stump so the neck is exposed and not too close to an edge.
5. To help hold the bird in place, we have positioned a foot onto the slack at top of the bag (but not stepping on the bird).

6. Wait for a calm moment, and then give a firm, assertive chop. This is not always easy, but it is a very quick and therefore more humane death than other options out there.
7. After the head is gone, the bird will still have some movement. Try to position the bag to drain if you can by hanging it.

There are other ways you can set up for killing, but the above two options are what we've done and what are most common. Be strong, and remember your goals.

Once the bird is still and drained, you can move straight to scalding and plucking.

Scalding

You will need your scalding water to be around 155-160 degrees Fahrenheit. Get it heating up sooner rather than later, since you can always cool it down by adding cold water. See the earlier section about batch sizes for more on planning to scald.

To scald:
1. Make sure water is the right temperature, 155-160 degrees Fahrenheit.
2. Add a few squirts of dish soap to the water.
3. Submerge the entire bird in the hot water for a full minute. We do 2 birds at a time in our 3 gallon pot.
4. At the 30 second mark, move the birds around to ensure the soapy water is penetrating to the skin everywhere.
5. Lift the chickens out of the pot and let excess water drip off for a few seconds.

6. Don't forget to set the pot back on the stove to heat back up for the next batch, if you have another batch to process.

Plucking

If you have a plucker, follow the directions that came with it. For a barrel-style plucker, you're looking at about 30 seconds of spinning before the birds are clean. Sometimes they will come out with some feathers still attached. This could be from scalding water not being hot enough. You could scald again, or you could just pull the feathers out by hand.

To pluck by hand, your job is simple. Pull out all the feathers. Try different approaches and you'll find something that works for you. We tend to prefer pinching feathers between a thumb and forefinger and pulling against the direction the feathers are angled.

Evisceration

Now it is time to remove the bits you don't want to eat. As you do this more and more, you will become comfortable with the anatomy and get better at breaking things down. But take your time at first; there is no need to rush. And you don't want to make things any messier than you need to at this stage.

Get some bowls or other vessels set up and make sure your knife is sharp. You'll want to have somewhere to toss bits you are discarding, and you'll be setting some things aside, like necks, feet, and organs.

DINNER FROM SCRATCH: HOW TO RAISE MEAT CHICKENS

1. Remove the head by cutting it off from the neck. You should be able to find a good spot between the vertebrae and not have to cut through bone.
2. Before you remove the neck, it is a good idea to use your knife to free the stuff around the bone and muscle. There is a crop and some tubing that you'll want to pull from the neck and not cut off (it's OK if you do, but leaving these intact helps keep things cleaner down the line). Also cut the skin away from the neck to make removal easier.
3. Now you can cut the neck out with your trusty knife, and set it aside for making stock or gravy.
4. Now turn the bird around and remove the feet— you can guide the blade between the joint fairly easily for a clean result. Save the feet for stock or interesting recipes if desired.
5. Now the guts. At the abdomen (where you would put stuffing into a turkey), pinch some of the skin near the top middle area and carefully pierce an opening with your knife. You want to be careful because puncturing an organ can lead to a mess. There is a little green thing in there (bile duct) that can release a staining liquid that could taint your meat and stink up the joint.
6. After piercing an opening, guide the knife to the side to make the opening bigger. You could also use your fingers to rip a larger opening.
7. Before you start yanking things out, you'll want to cut off the tail and free up the vent. The vent is where waste comes out, so take your time. Locate the tail and vent, and then cut into the lower tailbone area around the tail to free it from the rest of the body.
8. Now make sure there is no skin holding any of the guts in place, by making sure the opening you started with goes all the way around to where you cut the tail.

9. Reach in with your gloved hand and slide your knuckles all around the inside cavity to free things up. Take a gentle but firm hold of things and remove. If things aren't moving, go back to where you removed the neck and make sure the crop and tubing up there is freed up from the body and skin, which should allow most everything to slide right out.
10. You can find the kidney and heart among the things you just pulled out. The heart will be a sort of elongated ball, looking very much like your heart only small. The kidney will look like a very large kidney bean. Both will be fairly firm. Set these aside if desired.
11. Next you need to remove the lungs and, in the case of males, the "beans". The beans are their reproductive organ, and will be about the size of a black bean, only they will be off white. You can pop them out easily with a fingertip. The lungs may require extra effort—they will be red and pressed up against the ribcage. Work them out with your fingers.
12. Take a moment—you did the hard part!
13. Cut away any excess skin around the neck and abdomen area.
14. With a high powered spray, use your hose to wash out any debris from the cavity and outside of the bird.
15. Now it's just a matter of chilling the birds in some ice water while you finish up the others. Toss them in a cooler filled with ice and water and move on. Bravo!

Packaging & Freezing

If you are using shrink bags for your chickens, we recommend getting water heating up before you are ready to start bagging. You will want the water to be around 185-195 Fahrenheit for most shrink bags, and you'll need a pot big enough to accommodate a chicken fully submerged.

After the birds are processed and clean, you'll want to dry them off a bit. This makes it so there won't be as much ice build-up in the freezer. We take quart sized glass jars and prop the chickens up on them, with the top of the jar inserted into their cavity. You could also pat them dry with a towel.

Here's the step by step process for bagging with shrink bags:
1. Set up the birds to dry off. We prop them onto quart jars on top of an old towel.
2. Make sure the water is hot enough (185-195 degrees Fahrenheit).
3. Guide the chicken into the bag, "head" first. If you started legs first things might be a little more difficult, with legs and wings snagging the edge of the bag and generally annoying you.
4. Insert a length of narrow tubing (usually comes with the bags when you order them) into the chicken's cavity with the tube sticking out of the top of the bag. Cinch down the opening of the bag with a zip tie so it tightens around the tube. Remove the tube, but do not tighten the zip tie. This will leave a small hole for air to escape through as the bag shrinks.
5. Carefully dip the chicken bag into the hot water for 5 seconds, making sure the water touches all of the bird's surface area.

6. Remove from the water and cinch down the zip tie all the way.
7. Dry off the bags with a clean cloth or paper towels.
8. Weigh and label as desired. We use permanent marker and just write the weight in pounds right on the bag.
9. Throw the bird in the fridge or freezer and you are done!

No really, you are fully done! You have raised chickens for eating from start to finish. Congratulations!

Calculating Cost Per Pound

If you are interested in determining how much the endeavor of raising chickens costs, the most useful number you can determine is the cost per pound. This will give you a point of reference when comparing to other farmers and grocery store prices.

In order to calculate, you will need to record data as you raise your chickens:
- How much you spent on feed
- How much you spent on supplies
- You'll also need to estimate how much you spent on electricity, since you had to heat your brooder

Here's the most simple and direct way to calculate the cost per pound:

> cost per pound = ([total spent on feed] + [total spent on supplies] + [estimated electricity cost for brooding]) / [total pounds of meat yielded after all processing]

DINNER FROM SCRATCH: HOW TO RAISE MEAT CHICKENS

If you are going to be raising chickens for a number of years, you could calculate the cost for supplies differently. For example, if you spent $1500 on a plucker, you might not want to consider that as part of the expenses for just one flock of chickens, but rather you'd like to spread that out over however many years you think the plucker will be used. Here's how you can calculate that, in somewhat simple terms:

> cost per pound = ([total spent on feed] + [estimated electricity cost for brooding] + [total spent on supplies for just one year] + ([total spent on supplies that will be reused] / [number of years reused supplies estimated to last])) / [total pounds of meat yielded after all processing]

I hope you remember your order of operations! We do all of this with Google Sheets, by the way. Much easier to work out the formulas that way!

11 - MASTER SUPPLY LIST

- **General**
  - ☐ A poultry feeder sufficient for your number of birds (read product information). Avoid going too large, plastic preferred. Avoid tiny chick sized products.
  - ☐ Plastic Waterer that can support your number of birds. Avoid tiny chick sized products.

- **Phase I: Brooding**
  - ☐ Space for brooding
    - Secure from predators and pests
    - No wind
    - Ventilation / windows
    - Easy access for care and cleaning
    - Can hang heat lamps up to 4 feet high
    - Adaptable from small to large space for heat efficiency is a plus
    - 1 square foot per 2 chicks, plus space for feeders and waterers

- ☐ Heat lamps
- ☐ Chain to hang heat lamps
- ☐ S hooks for heat lamps
- ☐ Newspapers
- ☐ Pine shavings
- ☐ Bricks to raise feeder/waterer over time, or more chain and s hooks
- ☐ A good supply of broiler starter feed

- **Phase II: Pasture**
  - ☐ Movable pen(s) sized to support your flock comfortably
  - ☐ Tarp to cover around half of movable pen completely (all of back side, ⅔ of top, ⅔ of two sides
  - ☐ Rope or zip ties to secure tarp
  - ☐ Extra logs or blocks to weigh down movable pen's bottom flaps
  - ☐ Feeder(s)
  - ☐ Waterer(s)
  - ☐ Hose that can reach movable pens
  - ☐ Extra chain and s hooks
  - ☐ A good supply of broiler grower feed

- **Processing and storage**
  - ☐ Plucker (optional)—you can pluck by hand if you don't have a plucker, it just takes more time and effort

    If using a plucker, make sure you have any additional supplies for the plucker setup. Here's what we use with our barrel style plucker:

    - ☐ Y splitter for garden hoses if you only have one spigot
    - ☐ An extra hose to connect to the plucker

- ☐ Extension cord for plucker
- ☐ Garden hose that can reach your processing area
- ☐ Spray nozzle for garden hose
- ☐ Enough coolers to hold 10 birds
- ☐ Work table that can easily be washed and disinfected, like a foldable portable plastic table
- ☐ Trash barrel/bucket with trash bag
- ☐ Restraint cone (optional) mounted on something sturdy (a tree, saw horse, etc.)
  - If not using restraint you will need: chicken feed bag with corner cut out, an axe, and a large stump or wide log that you don't mind getting dirty or hitting with an axe.
- ☐ Bucket for draining chickens
- ☐ Rope for draining birds
- ☐ Large pot for scalding birds for plucking and shrinking bags
- ☐ Dish soap
- ☐ Candy thermometer
- ☐ Spray bleach solution
- ☐ Nitrile gloves
- ☐ Sharp paring knife or similar
- ☐ Plastic cutting board
- ☐ bowl
- ☐ Paper towels
- ☐ Work clothes and shoes that you aren't afraid to throw away if necessary
- ☐ Safety glasses
- ☐ Quart jars (optional, helpful for draining)
- ☐ Old towel or cloth you aren't afraid to throw away
- ☐ Bags to store chickens and any parts you are saving
- ☐ Source of entertainment (optional): we like to listen to podcasts while processing. In our

opinion, listening to music is a bad idea—you might end up forever associating the songs you listen to with the experience of processing, and the next time you hear the song you might flash right back to the work you will be doing with your chickens.

ABOUT THE AUTHOR

Brian is a passionate supporter of local food. On nights and weekends he works alongside his wife, Amy, to supply all of their family's maple syrup, chicken meat, eggs, and garlic. They seek out local sources for all other foods to supplement the other crops in their garden. They are also avid cooks and bakers.

By day, Brian is an experienced project manager, focused on improving process and achieving quality results with clear, simple procedures. He brings these skills to books like this one to help everyone achieve their own goals of living local and enjoying food deeply.

Visit his blog at https://ferrinbrookfarm.wordpress.com.

www.ingramcontent.com/pod-product-compliance
Lightning Source LLC
Chambersburg PA
CBHW020456220526
45464CB00002B/1013